OCT   1992

Plant-Eating Dinosaurs

PLANT-EATING DINOSAURS

by David B. Weishampel, Ph.D.

Illustrated by Brian Franczak

FRANKLIN WATTS
NEW YORK/LONDON/TORONTO/SYDNEY

This book is dedicated to my daughters:
Sarah Weishampel, who has kept me on my toes
when I've forgotten a dinosaur or two,
and Amy Weishampel, who is making sure
that I continue to learn them
at her amazingly quick pace

Library of Congress Cataloging-in-Publication Data

Weishampel, David B., 1952–
Plant-eating dinosaurs / by David B. Weishampel; illustrated by
Brian Franczak.
p.   cm.
Includes bibliographical references and index.
Summary: Examines the plant-eating dinosaurs and the
characteristics which enabled them to survive on that diet.
ISBN 0-531-11021-4
1. Dinosaurs—Juvenile literature.   2. Herbivores, Fossil—
Juvenile literature.   [1. Dinosaurs.   2. Herbivores, Fossil.]
I. Franczak, Brian.   II. Title.
QE862.D5W34   1992
567.9'7—dc20   91-16264 CIP AC

Contents

Introduction

Dinosaurs are indeed very special animals. Some of them were the largest land-living animals of all times. Others were quite small. Among the smaller creatures were meat-eaters called **coelurosaurs** (see-lur-o-sors), some of which evolved into the birds we see today. We also know that certain groups of dinosaurs made good parents and that others hunted in packs. All in all, dinosaurs have proved to be a very important and exciting group of animals that were successful for a very long time—nearly 160 million years. It was not until the extinction of these creatures that the kind of animals we see today, including our **primate** ancestors, were able to evolve.

Many of our favorite dinosaurs probably are of the meat-eating variety, like *Allosaurus* (al-o-sor-us) or even the larger *Tyrannosaurus* (ty-ran-o-sor-us). But did you know that there were many more kinds of dinosaurs that ate plants? In fact, **paleontologists** (scientists who study **fossils,** including dinosaur fossils) think that there were more than 300 different kinds of plant-eating dinosaurs. Some of these dinosaurs, such as *Stego-saurus* (steg-o-sor-us), *Brachiosaurus* (brack-i-o-sor-us),

**A scene of Jurassic herbivorous
and predatory dinosaurs**

and *Triceratops* (tri-sare-o-tops), may be familiar to you.
But others may not be so well-known.

All of the dinosaurs covered in this book are plant-
eaters. Like other plant-eating animals you know about—
such as horses, cows, giraffes, and zebras—the dinosaurs that fed on leaves, fruit, and stems had more than
a little difficulty in digesting this sort of food. The reason why is that **vertebrates** (animals with backbones)

lack certain substances, particularly important kinds of **enzymes** that can break up and digest the plant food in their stomachs. But plant-eating vertebrates have evolved certain tricks for digesting plant food. They have special chambers in the stomach, where small, one-celled organisms live that have the right substances to digest plants. This is how cows, horses, kangaroos, and giraffes manage to get nutrition out of the leaves, fruits, and grasses they eat. These plant-eaters also help the one-celled organisms that live in their stomachs with the digestion process by chewing their food thoroughly.

A scene of modern herbivorous and predatory mammals

This careful chewing chops up the leaves and grasses into smaller bits so that the one-celled organisms have an easier time digesting them. By chewing and by having special inhabitants in their stomachs, these animals have evolved a very important and useful way to feed on plants.

We don't know for certain if dinosaurs had fancy stomachs with one-celled organisms living in them,

10

because we do not have any fossil stomachs or one-celled organisms of these reptiles to examine. How then can we tell whether dinosaurs were good at feeding on plants? We can take a look at whether these animals chewed their food. Anything that tells us about an animal's ability to chop up leaves, stems, and fruits can also help tell us how the animal got its plant food ready for digestion in its stomach.

Geologic Time

Dinosaurs lived a long time ago, but not as long ago as the earliest animals and plants. The earliest dinosaurs evolved about 225 million years ago. Even though that is a long time ago, it's only one-twentieth as old as the earth. The last dinosaurs became extinct about 65 million years ago. That's one-seventieth as old as is the earth.

The age of dinosaurs is called the **Mesozoic** (mezz-o-zo-ic) Era. It was the time of warm and often somewhat dry climate nearly everywhere in the world. Even at the North and South poles, it was much warmer than it is today. It may not be too surprising then that dinosaurs lived just about everywhere on earth, even toward the poles. They also lived in different kinds of environments, from the mountains to the seashore. All of these environments were inhabited by dinosaurs throughout all of the divisions of the Mesozoic Era. These divisions are the **Triassic** (tri-ass-ic) Period, the **Jurassic** (jur-ass-ic) Period, and the **Cretaceous** (cree-tay-shus) Period. During the Triassic Period, from 248 million to 213 million years ago, the evolution of the first dinosaurs occurred. The Jurassic Period lasted from 213 million to

ERA	M.Y.A.*	PERIOD	
C E N O Z O I C	2	**QUATERNARY**	man
	65	**TERTIARY**	mammals
M E S O Z O I C	144	**CRETACEOUS**	Tyrannosaurus
	213	**JURASSIC**	sauropod dinosaurs
	248	**TRIASSIC**	first dinosaurs
P A L E O Z O I C	286	**PERMIAN**	mammal-like reptiles
	360	**CARBONIFEROUS**	amphibians
	408	**DEVONIAN**	armored fish
	440	**SILURIAN**	sea scorpions
	505	**ORDOVICIAN**	shellfish
	590	**CAMBRIAN**	trilobites
		PRECAMBRIAN	algae

*million years ago

Geologic time

144 million years ago. When it ended, the gigantic **sauropods** were the most common plant-eating dinosaurs. Finally, the Cretaceous Period began about 144 million years ago and ended 65 million years ago. During the Cretaceous Period, dinosaurs were most successful at being plant-eaters. It was also the time of the fearsome *Tyrannosaurus.*

One of the best ways to think about how long ago the time of dinosaurs was is to imagine all of **geologic time** as a single year. In the first moments of January 1, the earth was created. Life on earth evolved at about 10:47 A.M. on April 15. Higher forms of animals evolved at about 6:07 P.M. on November 14, while dinosaurs evolved on December 12 at 6:00 P.M. These animals became extinct on Christmas Day at 5:51 P.M. Finally, human beings evolved sometime around 8:06 P.M. on December 30. This was not much before the close of this geological year.

The Dinosaurs That Ate Plants

The first dinosaurs were small, active meat-eaters. Plant-eating dinosaurs evolved from these early meat-eating dinosaurs shortly thereafter. There were two basic groups of plant-eaters. One group evolved a special kind of hip and jaw structure that enabled them to be successful at feeding on plants. These are classified as **ornithischians** (or-nith-isk-i-ans). They include **stegosaurs** (steg-o-sors), **ankylosaurs** (an-ky-lo-sors), **ornithopods** (or-nith-o-pods), **pachycephalosaurs** (pack-e-sef-al-o-sors), and **ceratopsians** (sare-a-tops-i-ans). The other kind of plant-eater represents part of the **saurischian** (sor-isk-i-an) group of dinosaurs. They include **sauropodomorphs** (sor-o-pod-o-morfs) and **segnosaurs** (seg-no-sors). These dinosaurs are also related to the meat-eating theropods like *Tyrannosaurus* and *Troodon* (troe-o-don).

SAURISCHIANS

Sauropodomorphs

You may think you don't know any dinosaurs in this group, but you really do. The largest of all dinosaurs,

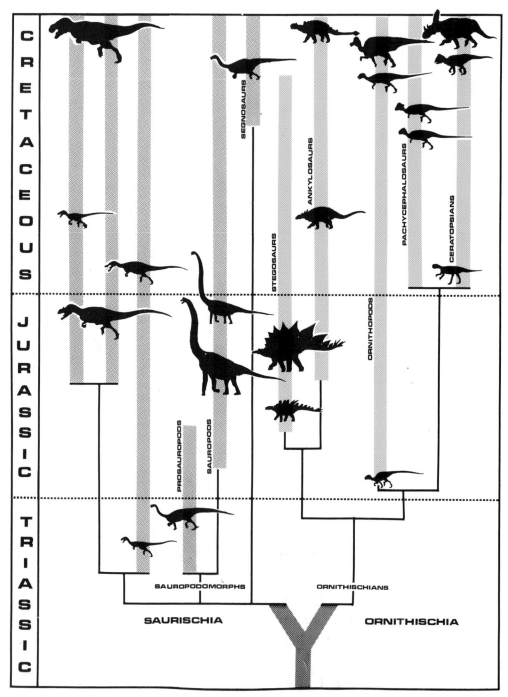

The evolution of dinosaurs
(dinosaurs not labeled are meat-eaters)

such as *Brachiosaurus, Diplodocus* (di-plod-o-cus), and *Apatosaurus* (a-pat-o-sor-us; the right name for *Brontosaurus* [bron-to-sor-us]) is found in this group of plant-eaters. They are called **sauropods** (sor-o-pods).

Just to give you an idea of how big these animals may have been, *Brachiosaurus* weighed approximately 60 tons, about as much as eight elephants. *Supersaurus* (su-per-sor-us) and *Ultrasaurus* (ul-tra-sor-us) were larger dinosaurs, each weighing perhaps as much as 100 tons. Also, a new kind of dinosaur, so far nicknamed *"Seismosaurus"* (size-mo-sor-us), perhaps weighing up to 600 tons, or as much as eighty elephants, has recently been found.

Most of the sauropods mentioned here lived in what is now the United States at the end of the Jurassic Period, about 150 million years ago. However, there were many more kinds of the gigantic plant-eaters. For example, a very long-necked sauropod named *Mamenchisaurus* (ma-mench-i-sor-us) lived about 150 million years ago in China. It had a relative living in what is now the Gobi Desert of Mongolia named *Opisthocoelicaudia* (o-pis-tho-seel-i-caud-ia). Giant sauropods lived just about everywhere in the world, including South America, Australia, India, and Europe.

All of these immense plant-eaters were very stoutly built, with thick legs, small heads, long necks, and sometimes very long tails. None of them had a very large brain, but apparently they had enough brainpower for their gigantic size. Sauropods as large as *Mamenchisaurus* probably never ran but walked at a slow, even pace.

The skeletons of these huge animals sometimes are found in great abundance, in areas almost like grave-

Brachiosaurus

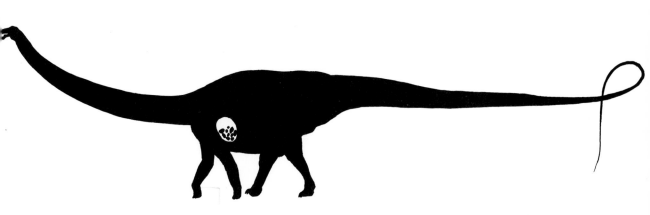

Sauropods broke up their food
with stomach stones.

yards. In other places, sauropod footprints are found in
great numbers of crisscrossing pathways. The footprints
come from both young and old sauropods. The trackways
and skeletons are the sort of evidence that scientists
can use to understand the lives of sauropod dinosaurs.
Both indicate that sauropods probably lived in large
groups composed of families and neighbors.

In addition to the sauropods, there were other kinds
of sauropodomorphs as well. The ancestors of the sauropods were called **prosauropods** (pro-sor-o-pods).
These plant-eaters lived at the end of the Triassic Period
and died out during the beginning of the Jurassic Period.
Prosauropods such as *Massospondylus* (mass-o-spond-
i-lus) were much smaller than the gigantic *Brachiosaurus*
and *Diplodocus.* However, like the enormous sauropods,
the prosauropods had long necks, stout bodies, and
long tails.

Plateosaurus (plat-e-o-sor-us) and other prosauropods probably walked mostly on their hind legs. Like

Mamenchisaurus

sauropods, the prosauropods lived in large herds. Quite a number of these animals are known to have been from Africa, but others were from North America, Europe, South America, and Asia.

Both prosauropods and sauropods ate plants. But they did so without actually chewing up the leaves and fruits that they were feeding on. From what we've already talked about, that may seem a bit strange. But prosauropods and sauropods developed something other than chewing that would break apart their food after they had swallowed it. They swallowed stones that were then stored in a muscular pocket, called the **gizzard,** that is part of the stomach. When prosauropods and sauropods swallowed leaves or other parts of a plant, these bits of food first went into the gizzard. The muscular gizzard then moved the rocks inside it to grind up the food. Although they didn't chew, prosauropods and sauropods evolved a very important and different way to feed on plants.

As peaceful plant-eaters, prosauropods and sauropods had a number of natural enemies. Small meat-eating dinosaurs such as *Syntarsus* (sin-tar-sus) fed on the much larger prosauropods. Later theropods, such as *Allosaurus* and *Ceratosaurus* (ser-at-o-sor-us), probably fed on sauropods, although attacking these gigantic plant-eaters must have been very difficult. As a result, *Allosaurus* and *Ceratosaurus* must have hunted sauropods in packs, in much the same way that wolves now hunt moose.

Segnosaurs

Segnosaurs (seg-no-sors) are a group of dinosaurs that have only recently been discovered by paleontologists. All of the segnosaurs lived during the last part of the Cretaceous Period in what is now Mongolia and Chi-

Massospondylus

Syntarsus

Massospondylus being attacked by the
smaller predatory **Syntarsus**

**Prosauropods used stomach stones
to break up their food.**

na. *Segnosaurus* (seg-no-sor-us) itself was more than 15 feet (4.5 m) long. It probably walked on all four of its legs, but when it wanted to run, it most likely used only its hind legs.

Segnosaurs had simple, triangular teeth. So far, no gizzard stones have been discovered with the skeletons of these kinds of dinosaurs. This means that segnosaurs probably ate only fleshy leaves and fruits. They must have fed on lots of them, because these dinosaurs had very big stomach regions!

ORNITHISCHIANS

Stegosaurs
The unusual dinosaur called *Stegosaurus* had a series of plates on its back and paired spikes at the tip of

Segnosaurus

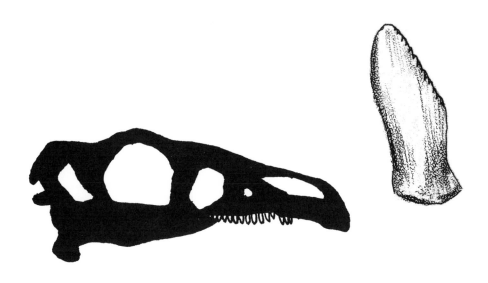

Segnosaur skull and tooth

its tail. It also had a small brain and may not have been very smart. Paleontologists have sometimes said that it also had a second brain between its hips. Some of these ideas about *Stegosaurus* are true and some are not.

The earliest known **stegosaur** is *Huayangosaurus* (hu-a-yan-go-sor-us) from the Jurassic Period in China. *Huayangosaurus* had two rows of very narrow plates from its shoulders to the tip of its tail, unlike *Stegosaurus*, which had a series of broad, triangular plates on its back. Other stegosaurs had spikes and plates along their backs and on their tails as well.

All stegosaurs ate plants. Most likely they chewed their food, because there is no indication that they had gizzard stones. But stegosaurs had simple teeth, which suggests that these animals ate soft kinds of plant food. Maybe they fed mostly on fruits or fleshy leaves, not on very tough leaves or twigs.

What about the theory that *Stegosaurus* had two

Stegosaurus

Allosaurus

Stegosaurus being attacked by an *Allosaurus*

Stegosaurs: their spinal cord,
plates, skull, and teeth

brains, one in its head and another between its hips? Although only about the size of a large walnut, the brain inside of *Stegosaurus*'s head was just about the size you might expect to find in a reptile of this size. Most animals that walk on land have some extra space between their hips for the nerves that control the muscles of the back legs. This includes humans as well. Maybe this area in *Stegosaurus* was larger because the dinosaur used its back legs and tail more than other kinds of animals. A

27

Huayangosaurus

large nerve center for the hind legs and tail may be what is meant by a second brain.

Like the sauropods that lived at the same time, stegosaurs had plenty of enemies among the meat-eating theropods. When the fearsome *Allosaurus* was not trying to make a meal of sauropod, it may have attacked a *Stegosaurus.* The spikes at the end of the tail of *Stegosaurus* would certainly have provided some defense, but what were those plates along its back used for? Paleontologists have often thought that the plates too would have been useful for protecting *Stegosaurus* against an *Allosaurus* attack. But it may be that they were also used like an air conditioner for cooling or a solar panel for heating. There are many holes and channels in each plate, and these were probably filled with blood vessels. A wide and flat plate with lots of blood flowing through it would make an excellent sort of air conditioner when *Stegosaurus* was a little too hot, or a solar heater when it was a little too cold.

In their day, stegosaurs were very successful. They thrived across the globe, especially in what is now China. These plated plant-eaters decreased in number and kind in the early part of the Cretaceous Period. And by the time the rest of the dinosaurs died off at the end of this period, not a single one was still alive.

Ankylosaurs

Ankylosaurs look like the armadillos of the Mesozoic Era. Like stegosaurs, ankylosaurs ate lots of plants. Since these armored dinosaurs had no gizzard stones to grind their food, they probably chewed what they ate. Their teeth were relatively simple and shaped like triangles, so their food probably consisted of the soft parts of plants, such as fruit and fleshy leaves.

These massive dinosaurs had a bony shield across their backs that protected them from being eaten. Nearly all of these animals had spikes of bone that stuck out from the sides and top of the shield of armor. Some even had a bony club at the end of the tail. Their spikes and bony tail clubs would have helped these armored dinosaurs defend themselves. For example, the fearsome *Tyrannosaurus, Albertosaurus* (al-bert-o-sor-us), and *Tarbosaurus* (tar-bo-sor-us) would have made a quick meal of an ankylosaur were it not for its armor. The small but equally ferocious meat-eating *Dromaeosaurus* (drome-ay-o-sor-us) and *Velociraptor* (vel-o-sir-apt-or) were also on the prowl for an ankylosaur meal.

Although they lived throughout most of the Mesozoic Era, there were more kinds of ankylosaurs at the end of the Cretaceous Period than at any other time. The most completely known of these armored dinosaurs is *Euoplocephalus* (yew-op-lo-sef-a-lus) from North America, but there were many others. *Sauropelta* (sor-o-pel-ta) also lived in North America, while *Shamosaurus* (sham-o-sor-us) lived in Asia, *Struthiosaurus* (struth-i-o-sor-us) was found in Europe, and *Minmi* (min-mee) in Australia. Like all dinosaurs, ankylosaurs became extinct at the very tail end of the Mesozoic Era.

Ornithopods

The name **ornithopod** (or-ni-tho-pod) means bird foot, but this name doesn't give much of an idea about what these dinosaurs looked like or how they lived. There are a great many different kinds of ornithopods. We'll be discussing a few of the more important kinds.

The earliest ornithopods were animals like *Heterodontosaurus* (het-er-o-dont-o-sor-us). *Heterodontosaurus* lived during the early part of the Jurassic Period in

Euoplocephalus

Ankylosaur skull and tooth

southern Africa, along with similar dinosaurs, like *Abrictosaurus* (a-brick-to-sor-us). These animals were about 6 feet (1.8 m) long and weighed between 50 and 100 pounds (22.5–45 kg). *Heterodontosaurus* walked and ran on its hind legs. Its front legs or forefeet were used to grasp branches and bits of food.

Heterodontosaurus and its relatives were among the earliest of plant-chewers. Large teeth lined the jaws. These were used to grind leaves and perhaps fruit into small bits. Muscular cheeks kept food from falling out of the sides of the mouth. In front of the grinding teeth were large tusks that probably protruded outside the lips of *Heterodontosaurus*. These ferocious-looking tusks were often used as weapons for protection against enemies like the meat-eating *Syntarsus*. They also may have been used against attacks by other males or for showing off between males and females.

The second big ornithopod group is made up of **hyp-**

Sauropelta

Deinonychus

Sauropelta being attacked by **Deinonychus**

silophodontids (hip-sill-o-foe-don-tids). Just about all of these dinosaurs were small (about 100 pounds [45 kg]) and fast runners. *Hypsilophodon* (hip-sill-o-foe-don) is probably the best-known member of this group. It lived in what is now England during the early part of the Cretaceous Period. There are many other kinds of hypsilophodontids as well. Some, like *Yandusaurus* (yan-du-sor-us), come from China, while others, like *Othnielia* (oth-neel-ia), *Thescelosaurus* (thess-kell-o-sor-us), and *Orodromeus* (or-o-drome-ee-us), are found in North America. Others come from Australia and even Antarctica.

Because they lived all over the world for a very long time, hypsilophodontids had many different kinds of enemies. *Yandusaurus* feared *Gasosaurus* (gas-o-sor-us),

Two *Heterodontosaurus* males
threatening each other

Othnielia trembled over *Ornitholestes* (or-nith-o-less-teez) and *Allosaurus,* and *Orodromeus* avoided *Troodon* at all costs. It was very important for these plant-eaters to be as fast on their feet as they could, in order to out-run their predatory enemies.

All of these ornithopods were great chewers. They certainly had strong chewing teeth, but they also evolved a special kind of jaw setup that we might find a little peculiar. Dinosaurs like *Hypsilophodon* and *Yandusaurus* developed a very loose skull that allowed them to chew sideways, in somewhat the same way that horses and camels do. By evolving these kinds of jaws, hypsilopho-dontids became very successful plant-eating dinosaurs.

Heterodontosaurid skull and tooth

The last big group of ornithopods, the iguanodontians (ig-wan-o-dont-i-ans), also had this unusual set of jaws. They were able to chew plant food as well as, or even better than, the hypsilophodontids. Iguanodontians most likely ate plants that were pretty tough and fibrous. Perhaps because of their special kind of chewing system, these animals became quite large. *Iguanodon* (ig-wan-o-don) lived in what is now Europe during the early part of the Cretaceous Period. It grew as large as 30 feet (9 m) in length and weighed as much as 5 tons. Other early iguanodontians, such as *Tenontosaurus* (ten-on-to-sor-us) and *Ouranosaurus* (oo-ran-o-sor-us), were equally large. All early iguanodontians had lots of enemies. *Tenontosaurus* feared the ruthless *Microvenator* (mi-cro-ven-ate-or) and *Deinonychus* (dine-on-y-cus), while *Iguanodon* fled from the ferocious *Megalosaurus* (megal-o-sor-us) and *Baryonyx* (barry-on-ix).

Of all iguanodontians, the duck-billed dinosaurs, or **hadrosaurids** (had-ro-sor-ids), are probably the best known. More hadrosaurid fossils have been found than

35

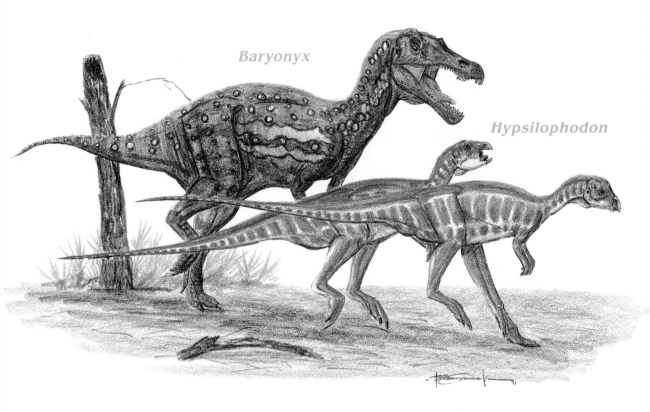

Baryonyx

Hypsilophodon

**Two *Hypsilophodon* being attacked
by a *Baryonyx***

probably than fossils for any other kind of dinosaur, and many skeletons are complete from head to tail. These dinosaurs lived mostly in what is now North America and Asia during the late part of the Cretaceous Period. But a few kinds lived in what is now Europe and South America. All hadrosaurids had a snout shaped somewhat like a duck's bill. Behind this snout, the jaws were lined with a tremendous number of teeth. Sometimes as many as 300 teeth were used for chewing at a single time!

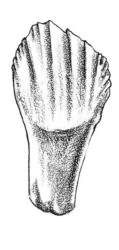

Hypsilophodontid skull and iguanodontian tooth

In addition, hadrosaurids had many different head shapes. *Edmontosaurus* (ed-mont-o-sor-us) and *Gryposaurus* (gry-po-sor-us) were flat-headed hadrosaurids. Both *Brachylophosaurus* (brack-ee-lofe-o-sor-us) and *Maiasaura* (my-a-sor-a) had a broad chest or shield across the top of the head. In a similar way, *Prosaurolophus* (pro-so-rol-o-fus) and *Saurolophus* (so-rol-o-fus) each had either a bony bump or a long bony spike extending from the top of the head. Perhaps the strangest creatures were *Corythosaurus* (cor-ith-o-sor-us), *Hypacrosaurus* (hi-pack-ro-sor-us), and *Parasaurolophus* (pear-a-so-rol-o-fus). There was a hollow crest extending above the top of their heads.

The purpose of each of these head ornaments has puzzled paleontologists for a long time. The latest and probably the best theory is that they were there for display purposes—that is, to show other dinosaurs what kind they were. The hollow crests of *Parasaurolophus*,

Corythosaurus

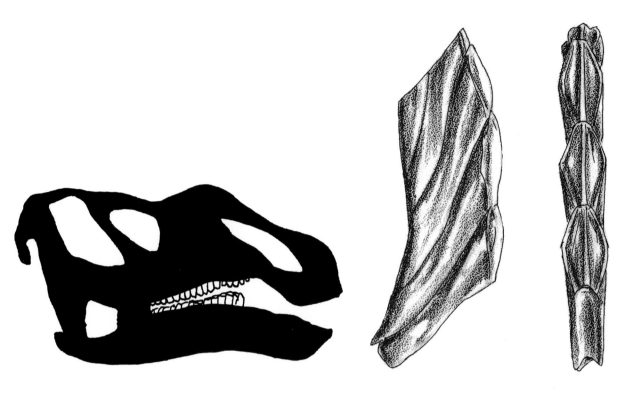

Hadrosaurid skull and teeth

Hypacrosaurus, and *Corythosaurus* were very special in this regard. They were not only showy, but they also probably helped the animals make a unique kind of sound in much the same way as a trombone or tuba works. These hollow crests allowed the dinosaurs to make very low notes. Such notes would have been useful as calls across great distances and in a variety of habitats, from forests to open areas. This kind of call allowed the dinosaurs to communicate among themselves.

Communication is a new and important aspect of dinosaur behavior. Adults could communicate with each other and with their young. Paleontologists know about

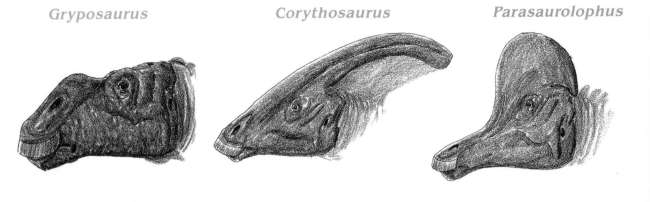

Gryposaurus   Corythosaurus   Parasaurolophus

Lambeosaurus   Saurolophus

Hadrosaurids with their different-shaped heads

the young of hadrosaurids (and other kinds of dinosaurs) because they have begun finding and studying eggs, nests, and bones of hatchlings. These fossils have taught us that hatchlings of hadrosaurids stayed in their nests for a long time. While there, the parents brought food for them. Perhaps the hatchlings climbed out of their nests for some exercise. More likely, they were helpless and not able to fend for themselves until they finally had grown up and left the nest for good.

Because hatchling hadrosaurids were defenseless in their nests, they were really at risk from their enemies. Packs of *Troodon, Dromaeosaurus,* or *Saurornitholestes* (sor-or-nith-o-lest-eez) could easily have raced into the dinosaur nesting area and picked out an unsuspecting hatchling for lunch. A giant *Tyrannosaurus* or *Alberto-saurus* could also have made an easy meal of a couple of hatchlings living in the same nest. It is probably a good thing that dinosaurs nested together in the same area, sort of like a colony. This way the adults could protect their young from being attacked by predatory dinosaurs.

Heterodontosaurus and its relatives all died off some-time during the early part of the Jurassic Period. Hypsil-ophodontids lasted until the end of the reign of dino-saurs, as did the iguanodontians. Then they, too, died out at the close of the Cretaceous Period, 65 million years ago.

Pachycephalosaurs

Pachycephalosaurs are among the most amazing dinosaurs. The extraordinary thing about them was their head. In all pachycephalosaurs, the top of the head was greatly thickened. In *Pachycephalosaurus* (pack-ee-sef-al-o-sor-us) itself, the top of the skull was made up of bone 10 inches (25 cm) thick. Because of their thick heads, dinosaurs were known as dome-heads. Many of them also had knobby horns on their noses and the backs of their skulls.

Pachycephalosaurs come in different shapes and sizes. The largest, *Pachycephalosaurus,* was about 25 feet (7.5 m) long, but most were much smaller. For ex-ample, *Stegoceras* (steg-a-ser-as) and *Prenocephale* (preen-o-sef-al-ee) were dome-headed dinosaurs that

Two *Stegoceras* males having
a head-butting contest

grew only to about 6 to 8 feet (1.8–3 m) in length. Other pachycephalosaurs were quite different from *Pachycephalosaurus* and *Prenocephale.* Instead of having large domed heads, *Homalocephale* (home-al-o-sef-al-ee) and *Goyocephale* (goy-o-sef-al-ee) had somewhat flattened skulls.

The domes of the pachycephalosaur dinosaurs were used during head-butting, the way big-horn sheep and goats protect themselves with their horns. Imagine a pair of *Stegoceras* or *Prenocephale* running at each other, with their heads down and domes facing front. After several collisions, one of the butters would be the winner of the contest and the loser would slink away. Maybe these dinosaurs had head-butting contests during mating season or when defending a plot of land or territory. Whatever the reason, the domes of these dome-headed dinosaurs were well designed for head-to-head combat.

Pachycephalosaurs had teeth very similar to those of stegosaurs and ankylosaurs. These simple triangular teeth couldn't have done much heavy-duty chewing. Nor did these dinosaurs have gizzard stones to help break up food in the stomach. As a result, pachycephalosaurus probably fed on fleshy fruits, leaves, and stems.

The dome-headed dinosaurs had many enemies, both large and small. Among the greatest threats were *Tyrannosaurus, Albertosaurus,* and *Daspletosaurus* (dazz-plete-o-sor-us), but many pachycephalosaurs probably had more to fear from the packs of smaller predators, such as the fierce *Dromaeosaurus* and *Troodon.*

Pachycephalosaurs lived mainly in North America and Asia, but some have been found in Europe and Africa. All of them lived toward the end of the Cretaceous Period. Then they, too, became extinct about 65 million years ago.

Ceratopsians

Ceratopsians are the horned dinosaurs. Probably the best-known of these horned dinosaurs is *Triceratops.* But did you know that there were more than twenty-five other kinds of ceratopsians? The earliest of these dinosaurs was *Psittacosaurus* (sit-ac-o-sor-us). These parrot-beaked plant-eaters were small (about 6 feet [1.8 m] long, weighing 100 pounds [45 kg]) and lived during the early

Homalocephale

part of the Cretaceous Period in what is now China, Mongolia, and the eastern parts of the Soviet Union. *Psittacosaurus* was a special kind of plant-eating dinosaur because it had teeth that were used in chewing *and* a muscular gizzard filled with stones! We do not know what kinds of plants it ate, but they were certainly well broken up before they got to the stomach.

Somewhat later, **protoceratopsids** (pro-to-sare-a-top-sids) evolved during the late part of the Cretaceous Period. These kinds of ceratopsians were a little bit larger than *Psittacosaurus* (protoceratopsids weighed about 200 to 300 pounds [90–135 kg]) and had evolved a small

 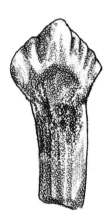

Pachycephalosaur skull and tooth

shield or frill at the back of the head. This frill protected the neck from attack by predatory dinosaurs and may also have been used to show off to one another. Males may have also used these frills to intimidate each other or to impress females during mating season.

The best known protoceratopsid is probably *Protoceratops* (pro-to-ser-a-tops) itself, from the Gobi Desert of Mongolia and northern China. Other protoceratopsids, such as *Leptoceratops* (lept-o-ser-a-tops) and *Montanoceratops* (mon-tan-o-sare-a-tops), lived in North America. Like *Psittacosaurus,* these protoceratopsids ate plants. However, these plant-eaters did not have gizzards, so they relied completely on chewing to break up their food.

Ceratopsids (sare-a-top-sids) were the last group of ceratopsians to evolve. They were very large and lumbering dinosaurs, all of which lived in what is now North America during the last part of the Cretaceous Period.

Psittacosaurus

Nanotyrannus

Leptoceratops

**A *Leptoceratops* being attacked
by a *Nanotyrannus***

Triceratops may be the best known of these ceratopsids, but there were many other forms. *Styracosaurus* (sty-rack-o-sor-us), *Centrosaurus* (sen-tro-sor-us), and *Pachyrhinosaurus* (pack-ee-rine-o-sor-us) all had somewhat short frills, extending only partway over the shoulders.

Chasmosaurus

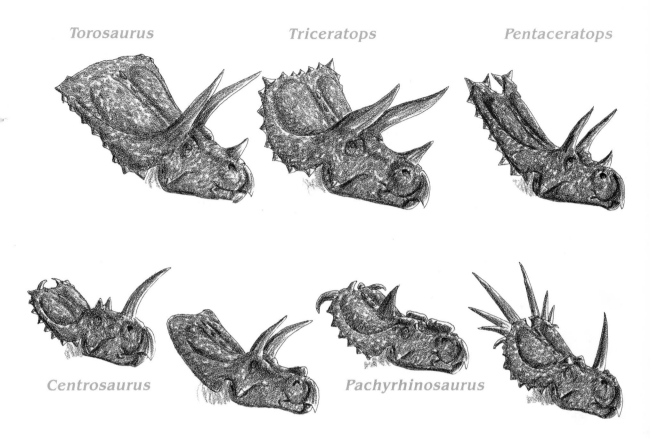

Torosaurus   *Triceratops*   *Pentaceratops*

Centrosaurus   *Pachyrhinosaurus*

Arrhinoceratops   *Styracosaurus*

Ceratopsids with their different-shaped heads

Both *Styracosaurus* and *Centrosaurus* had a very large horn over the nose and two short horns, one over each eye. In addition, *Styracosaurus* had many horns sprouting from the edge of the frill. *Pachyrhinosaurus*, on the other hand, had the most unusual face. Instead of a horn over the nose, it had a large and rough bony pad.

Other ceratopsids, such as *Chasmosaurus* (cazz-moe-sor-us), *Torosaurus* (tor-o-sor-us), and *Pentacera-tops* (pent-a-ser-a-tops), had very long frills extending over the shoulders. In addition, a long horn grew over

Ceratopsian skull and teeth

each eye and a prominent horn extended from the top of the nose.

The frills and horns in ceratopsids probably served mainly for show, much as they did with protoceratop-sids. Males would try to dominate each other by showing off their frills and swinging their horns around. Maybe they even used these horns in combat with each other. Of course, these horns and frills also protected each animal from its enemies. And there were lots of enemies, including the large and ferocious *Tyrannosaurus, Alberto-saurus,* and *Daspletosaurus.* Packs of the smaller predators such as *Dromaeosaurus, Troodon,* and *Saurornitho-*

lestes may have threatened these large plant-eaters as well.

Ceratopsids were among the most numerous plant-eaters during their time. Like protoceratopsids, their parrotlike beaks snapped up leaves and branches, while their scissorlike teeth probably sliced and diced them before they were swallowed.

Psittacosaurus became extinct by the end of the early part of the Cretaceous Period, but protoceratopsids and ceratopsids were around much longer. Both groups lived throughout the end of the Cretaceous Period, only to become extinct, along with all other dinosaurs, about 65 million years ago.

Conclusion

All of the plant-eating dinosaurs discussed in this book have long since vanished, but when they were alive they were the largest and most numerous plant-eating animals that lived on land. Each group was special in the way it lived and evolved. Plant-eaters with simple teeth, such as stegosaurs, pachycephalosaurs, and ankylosaurs, fed on fruits and fleshy leaves, while ornithopods and ceratopsians ate tough vegetation, using their flat, grinding teeth and special jaw mechanisms. Finally, there were the prosauropods and sauropods that had gizzard stones for grinding plant food in their stomachs.

Plant-eating dinosaurs thrived for over 160 million years. That's most of the Mesozoic Era. But 65 million years ago, they all became extinct. Some scientists think that dinosaurs became extinct because they could not survive the slowly changing climate and its effect on their habitats. Other scientists think that dinosaurs died out because an asteroid collided with the earth. This impact would have been followed by terrible changes in the dinosaurs' environment, such as long periods of darkness, severe cold spells, and the dying-off of important groups of plants. Whether it was a slow change in climate or rapid changes caused by an asteroid that forced them to become extinct, dinosaurs didn't survive beyond the end of the Cretaceous Period.

The extinction of both the plant-eating and the meat-eating dinosaurs may have been a lucky break for our own ancestors. Mammals evolved when the dinosaurs did and lived throughout the age of dinosaurs. But all of these mammals were small, none larger than a cat. Many probably lived in trees and shrubs. They also may have been active only at night, much as opossums and raccoons are today, in order to avoid the fearsome dinosaurs that roamed by day. But when the threat of the dinosaurs vanished, different groups of mammals came out of hiding and evolved different ways of living. They grew bigger and developed different kinds of eating styles and ways of walking and running. These new kinds of plant-eating and meat-eating mammals evolved over the next 65 million years—to the present day. Among these new kinds of mammals were our very early ancestors. In a way, we owe our life—and very presence here—to the disappearance of the dinosaurs.

Glossary

(The individual dinosaurs discussed in the book are not defined in the glossary.)

Ankylosaur (an-ky-lo-sor): The group of armored dinosaurs, classified with ornithopods, stegosaurs, pachycephalosaurs, and ceratopsians as ornithischian dinosaurs.

Ceratopsian (sare-a-tops-i-an): The group made up of horned dinosaurs, which together with stegosaurs, ankylosaurs, ornithopods, and pachycephalosaurs, are classified as ornithischian dinosaurs.

Ceratopsid (sare-a-top-sid): The group of ceratopsians that includes *Triceratops* and *Centrosaurus*, among others. These animals often have bony frills extending from the backs of their heads and horns on their faces.

Cretaceous (cree-tay-shus): The last period of the Mesozoic Era. The Cretaceous Period lasted from 144 million to 65 million years ago.

Enzyme (en-zyme): The group of chemicals that helps

to make changes in different kinds of substances, such as breaking down the cell walls of plants.

Fossil (fos-sil): Hardened remains or traces of plant or animal life preserved in rock in the earth's crust.

Geologic time: (ge-ol-oj-ic tyme): The succession of time since the origin of the earth to the present.

Gizzard (giz-erd): The thick muscular part of the stomach of many birds and reptiles.

Hadrosaurid (had-ro-sor-id): The duck-billed dinosaurs. These animals lived during the last part of the Cretaceous Period.

Hypsilophodontid (hip-sill-o-foe-don-tid): One of the groups of ornithopods. Hypsilophodontids were small and fast-running. The group includes *Hypsilophodon*, *Orodromeus*, and *Yandusaurus*, among others.

Iguanodontian (ig-wan-o-dont-i-an): One of the groups of ornithopods. Iguanodontians were large and include *Iguanodon*, *Ouranosaurus*, and hadrosaurids, among others.

Jurassic (jur-ass-ic): The middle period of the Mesozoic Era. Lasting from 213 million to 144 million years ago, the Jurassic Period is best known for its large sauropod dinosaurs.

Mesozoic (mez-o-zo-ic): The middle era (248 million to 65 million years ago) of what is called the Phanerozoic Eon (the time of complex life), which comprises the past 570 million years.

Ornithischian (or-nith-isk-i-an): One of the two major groups of dinosaurs. Ornithischians consist of stegosaurs, ankylosaurs, ornithopods, pachycephalosaurs, and ceratopsians.

Ornithopod (or-nith-o-pod): The group of bird-footed dinosaurs. Ornithopods include heterodontosaurids, hypsilophodontids, and iguanodontians. Ornithopods are one of the groups comprising ornithischian dinosaurs.

Pachycephalosaur (pack-e-sef-al-o-sor): The group of dome-headed dinosaurs. Pachycephalosaurs combine with ornithopods, stegosaurs, ceratopsians, and ankylosaurs to make up the ornithischian dinosaurs.

Paleontologist (pail-ee-on-tol-o-jist): A scientist who studies paleontology.

Paleontology (pail-ee-on-tol-o-jee): The science of prehistoric life.

Primate (pry-mate): The group of mammals that includes prosimians, monkeys, and apes (including humans).

Prosauropod (pro-sor-o-pod): The group of sauropodomorphs that includes *Plateosaurus* and *Massospondylus*, among others.

Protoceratopsid (pro-to-sare-a-top-sid): One of the groups of ceratopsian dinosaurs. Protoceratopsids include *Protoceratops*, *Montanoceratops*, and *Leptoceratops*, among others.

Saurischian: (sor-isk-i-an): One of the two main groups making up dinosaurs (the other is ornithischian dinosaurs). Saurischians consist of theropods, sauropodomorphs, and segnosaurs.

Sauropod (sor-o-pod): The largest of all land-living animals. Sauropods include *Apatosaurus*, *Diplodocus*, and *Supersaurus*, among others.

Sauropodomorph (sor-o-pod-o-morf): One of the groups of saurischian dinosaurs. Sauropodomorphs include prosauropods and sauropods.

Segnosaur (seg-no-sor): One of the groups of saurischian dinosaurs and the newest group of dinosaurs to be discovered.

Stegosaur (steg-o-sor): One of the groups of ornithischian dinosaurs. Stegosaurs had either plates or spikes (or both) along their backs.

Theropod (ther-o-pod): The group of meat-eating dinosaurs. Theropods are grouped with sauropodomorphs and segnosaurs to form the saurischians.

Triassic (tri-ass-ic): The first period in the Mesozoic Era. It began about 248 million years ago and ended about 213 million years ago.

Vertebrate: (vert-a-brate): The group of animals with backbones.

For Further Reading

Booth, Jerry. *The Big Beast Book: Dinosaurs and How They Got That Way.* Boston: Little, Brown, 1988.

Crenson, Victoria. *Discovering Dinosaurs: An Up-to-Date Guide Including the Newest Theories.* Los Angeles: Price Stern, 1988.

Horner, John R., and James Gorman. *A Dinosaur Grows Up.* Philadelphia: Running Press, 1989.

Norman, David. *The Illustrated Dinosaur Field Guide.* New York: Crown Press, 1989.

Index

Page numbers in *italics* indicate illustrations.

About the Author

David Weishampel is an associate professor of cell biology and anatomy at the Johns Hopkins University School of Medicine in Baltimore, Maryland. Since entering the discipline of vertebrate paleontology, Professor Weishampel has been involved in fieldwork in central Utah, Kansas, West Virginia, and Alberta (Canada). Currently, he is conducting fieldwork in northwestern Montana. The work includes studies of dinosaur family structure.

Professor Weishampel's two daughters, ages ten and five, have been out in the field since the time they were infants, and are enthusiastic dinosaur experts in their own right.